# *Quantum Ecology*

## Energy Structure and its Analysis

*László Orlóci FRSC*
*Western University, London, Canada*

2nd enlarged edition

**SCADA PUBLISHING – LONDON - CANADA**

**Refer to this book:**

Orlóci, L. 2015. Quantum Ecology. Energy structure and its analysis. 2nd enlarged edition: SCADA Publishing, Canada. Online Edition: https://createspace.com/4406077

**Look for these books:**

Orlóci, L. 2014. The vegetation process. A holistic study of long-term community energetics in East Beringia. SCADA Publishing, Canada. Online Edition: https://createspace.com/4760258

Orlóci, L. 2013. Quantum analysis of primary succession. The energy structure of a vegetation chronosere in Hawai'i Volcanoes National Park. SCADA Publishing, Canada. Online Edition: https://createspace.com/4452597

Orlóci, L. 2013. On the Energy Structure of Natural vegetation. In search for community governance rules. SCADA Publishing, Canada. Enlarged Online Edition: https://createspace.com/4153484

Orlóci, L. 2012. Self-organisation and Mediated Transience in Plant Communities. SCADA Publishing, Canada. Enlarged Online Edition: https://createspace.com/3585127

Orlóci, L. 2011. Statistical Ecology. The quantitative exploration of nature to reveal the unexpected. SCADA Publishing, Canada. Online Edition: https://createspace.com/3476529
Orlóci. L. 2011. Problem flexible computing in statistical ecology. SCADA Publishing, Canada. Online Edition: https://www.createspace.com/3574792

Orlóci, L. 2012. Statistical multiscaling in dynamic ecology. Probing the long-term vegetation process for patterns of parameter oscillations. SCADA Publishing, Canada. Online Edition: https://createspace.com/3830594

ISBN-13: 9781492183297
ISBN-10: 1492183296

**2nd enlarged edition V-2016-07-31**

Front cover: Pōhuehue (Ipomoea pes-caprae subsp. Brasiliensis) Kaewaula Beach, Oahu

Look for information: https://sites.google.com/site/statisticalecology/

Scada
Publishing

*"Alle Gestalten sind ähnlich, und keine gleichetder andern;*
*und so deutet das Chor auf ein geheimes Gesetz, ...."*

Johann Wolfgang von Goethe (1798): "Die Metamorphose der Pflanzen "

In free translation:

"All [plant] forms are similar, but none like any other;
and so the ensemble [community] interprets a hidden law ..."

Márta 2014, Lahaina

My thanks are due to:
Márta Mihály BSF, DFE for the Coquihalla data set and attentive
reading of the manuscript
Valério De Patta Pillar PHD for the Campos data
Carolina Blanco for selecting the data for quantum analysis

**Keywords**: Assembly; Brazil; Campos; Complex; Coquihalla; Disassembly; Diversity; Dynamics; Energy; Entropy; Environment; Evolution; Floodplain; Governance; Grazing; Hierarchy; Metacommunity; Phylogenetic tree; Potential energy; Probability

# Table of contents

# About the book

Quantum Ecology is the science in which ecology joins forces with quantum theory to create a holistic approach in energy studies. The infusion of quantum theoretical principles into the ecological approach enables the investigative focus to shift from calorific flow in the ecosystem to the potential energy structure in the metacommunity[1] and its intrinsic or extrinsic determinants.

At the core of Quantum Ecology is the energy equation $E = E_{Phy} + E_{Env} + E_{Rnd}$. It is consistent with ecological theory which holds that the potential energy structure of a metacommunity is the sum of the energy footprints of phylogeny, current environmental mediation, and emergent effects. These three are weaved by nature into an intrinsic convolution.

The energy equation's quantum theoretical underpinnings are found in Max Planck (1901) postulates concerning energy-based entropy (EBE). EBE is defined by the simple equation $E = \ln C$. Symbol $C$ stands for the number of ways a metacommunity can be assembled under the rule of chance, from $n$ given taxa whose total performance in the metacommunity is $T$. Since $C$ depends on $n$ and $T$, EBE is a holistic scalar of the potential energy level in the metacommunity.

---

[1] An arbitrarily delineated piece of the vegetation. Its elements are plant taxa whose assemblage is endowed with the functionality of self-organization.

Notwithstanding the fact that EBE was introduced as an alternative scalar for the energy level in nano-scale resonator complexes[2], under which the link of EBE to the probability $P=C^{-1}$ is proven, the book argues that this link is extendable to the metacommunity under specific regularity conditions.[3]

Statistical techniques are developed on the pages of the book. Their principle objective is the isolation of the three energy footprints.

EBE's ability to be parameterised by conventional survey or experimental data gives breadth to the application of quantum analysis in all fields of Ecology.

Following the first printing of the book, the author has taken up the topic of ecological EBE analysis in several monographs. The insights gained through these are incorporated into the Book's revised and expanded text. The underlying theory, and modus operandi are discussed and examples are presented. The examples use survey type vegetation data from the Coastal Western Hemlock zone on the Lower Mainland of British Columbia and the Campos Sulinos Formation in Brazil's Rio Grande do Sul.

---

[2] In a typical case of plant ecology, the resonator complex is a vegetation metacommunity and the community elements, the plant taxa. The taxa are populations, recognised by traits of inheritance or function, are the resonators.

[3] <u>Postulate</u> a. EBE is an alternative parameter of the potential energy state in a nanoscale resonator complex, functioning under the Normal probability law. Under this postulate, potential energy E is linked to probability P which is $C^{-1}$. <u>Postulate</u> b. The energy units (not the energy quantities) are discrete, countable entities. This allows the measurement of potential energy level in the metacommunity in EBE terms on the basis of plant population or plant community performance.

# Introduction

## The vegetation's metacommunity

It is quite appropriate to begin our discourse by observing that we deal with a medium, *the vegetation*, which has functional connectedness from patch to biome. Therefore whatever technical arguments we put forward about the vegetation it must be anchored in the manageable unit, the vegetation stand. A more expressive term for "stand" is *metacommunity*.

Our definition of the "metacommunity" as a concrete measurable unit gains further clarity if we contrast it to the vegetation as the undefined patch or undefined biome object, or to the vegetation "type" which is an abstraction by virtue of being a collection or population of metacommunities. [4]

What the reader can see on the photograph in Figure 1 is a concrete unit, a metacommunity. Its extent as shown is limited. The size is not fixed. If I broadened or narrowed the

---

[4] Modern usage of the prefix "meta" is not quite uniform and may not even follow the original Greek which is the same as the Latin *post-*, expressing adjacency. Taxonomists imply this meaning when they express phylogenetic adjacency between two taxa, such as *Sequoia* and *Metasequoia*. I rather prefer the more frequent "X about X" form of usage. Accordingly the metacommunity, as a concrete unit, tells about the "vegetation", the real patch-to-biome unit.

field of view on my camera I could still say that the picture is the photographic image of a metacommunity. The real three-dimensional size matters, of course, but it has to be chosen to suit the purpose to be served.

Figure 1. Forest and grassland complex in the Campos Sulinos Biome of Brazil.

The site is typical for the Campos on the highlands in the state of Rio Grande do Sul, near the edge to the East and South in contiguity with the Atlantic Rainforest on the steep slopes of the escarpment. Long running ecological studies made the Campos Sulinos a well described biome[5].

I use a triplet of temporal relevés (record sets) of the biome, from the Eldorado research site of Professor V. De Patta Pillar on the lowland west of Porto Alegre. Dr. Carolina Blanco, a co-investigator in Pillar's grazing experiments, selected

[5] Pillar and Quadros (1997), Pillar, Müller, Castilhos and Jacques (2009), Pillar and Duarte (2010), Pillar, Blanco, Müller, Sosinski, Joner and Duarte (2013).

the actual data for quantum analysis.

To show broader versatility of the technique, I subject yet another set of metacommunities to quantum analysis. That set comes from the Coquihalla floodplain at Hope, British Columbia (Figure 2). The vegetation formation of the region is identified by Krajina (1959, Orlóci 1965) as a subzone of the Coastal Western Hemlock Zone of the Pacific Northwest. The metacommunities in this case are isolated 40 x 40 meter square area units systematically arranged on M. Mihály's belt transect across the floodplain (line A on the Google map in Figure 2)[6].

Figure 2. Google map of the M. Mihály transect site on the Coquihalla floodplain at Hope, British Columbia.

At the time of the survey (July 7-8, 1976) the flow in the river was uncontrolled, floods were seasonal events, and the site was isolated from the residential areas of Hope. The flow was torrential in May and June, several meters higher than the normal, the reaching on average 12 m³/s, following run-off from snow melt after warm days or rain with a roughly

[6]http://maps.google.com/maps/ms?ie=UTF&msa=0&msid=21662637730984531 3321.00049bf3254ab94ea225e

24 hour lag. The yield dropped to a mere average 2 m³/s from October to March. The river has a 740 km² watershed in the Cascade Mountains. The highest peak in the watershed is around 2000 m.

# The hierarchical relevé

Each metacommunity offers a variety of structural, functional and evolutionary traits for measurement. Since the states of the traits are shaped by the basic processes, historic and current, the descriptive scheme has to be adapted to the unique complexities of their measurement.

My *hierarchical relevé* does the job well. As applied, a relevé holds the records of two processes in one scheme: phylogeny and environmental mediation. These are my present choices for further scrutiny because of their unquestionable role in the shaping the EBE state of the metacommunity.

I construct hierarchical relevés in the manner of inverted phylogenetic trees (dendrograms) with measured records of taxon performance attached to the nodes. Figure 3 has three examples (RL1, RL2, RL3). We can interpret these as the records of metacommunities from three sites or alternatively of one metacommunity in three different states.

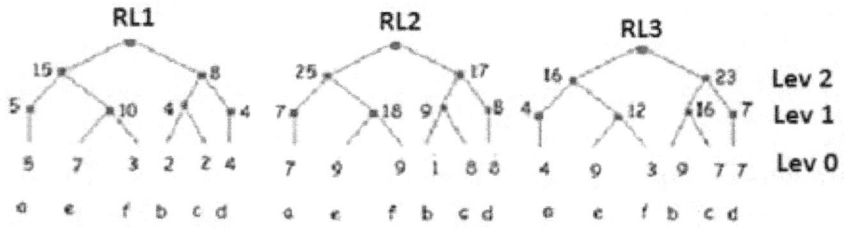

Figure 3. Hierarchical relevés.

As RL1, RL2 and RL3 are set up, six baseline taxa (a to f) are recorded. These are species in the examples fitted to a nested hierarchical system based on phylogenetic principles.

Having the baseline taxa defined as organismal species (Level 0), it is logical to define the higher units as Genera (Level 1) and Families (Level 2). In the examples to be presented the hierarchical levels go up to Order.

To the data analyst RL1, RL2 and RL3 are hierarchical record sets of an embedded kind. The elements in the baseline sets [5 7 3 2 2 4], [7 9 9 1 8 8], [4 9 3 9 7 7] are indivisible, but can be summed to form cumulants on higher hierarchical levels. The cumulants are the numbers placed at the nodes in Figure 3. Their sum is the same (23, 42, 39) on any level.

When we deal with species we accept that we are dealing with unique populations of plants which shared inheritance. Species are on the null level of the phylogenetic taxonomic system. Any node in the dendrogram identifies a point at which a higher level taxon has split into lower level taxa in the course of evolution. Not all splits may occur in the local flora. The hierarchical relevé is indeed local proxy for the true phylogenetic tree. [7]

---

[7] To answer arguments in specific cases is beyond my intention. I leave that discussion to the experts of evolutionary plant systematics. I was kindly warned that ranking taxa by their systematic status as in Figure 3 is quite arbitrary, just as much, as I see it, as any applicable plant systematics upon which the available field manuals are based, and from which I constructed my taxon ranking. This is notwithstanding the designation of this manuals as being based on phylogenetic principles. I wait for radically changing my hierarchical relevé scheme until I am handed a plant system based on pairwise divergence times of taxa covering the flora. Earth shaking changes are in the works which are making classical plant systematics appear increasingly obsolete. I see this clearly from János Podani's paradigm setting book "The evolution and classification of plants" published in Magyar by Elte Eötvös Kiadó, Budapest, 215.

# Propositions, principles

Modern Ecology is dated from the time of a major paradigm shift in the 19th Century from the naturalists' wonderment about Nature's work to the rules which govern Nature's wanders. I consider Anthon Kerner von Marilaun's 1863 doctrine of this kind of scientific merit regarding the governance of the vegetation process. Kerner has phrased the doctrine in his wonderful anecdotal style at length in "Das Pflanzenleben der Donau lander", but unmistakably the process he describes is what ecologists call "facilitation".

The heavy torrent of propositions kept inundating ecology ever since[8]. The ones of central interest to me, in this essay, are those which emerge when the proposition is successfully interfaced with a consistent analytical model. When this happens, the proposition becomes a guiding principle of ecological theory.

I should not dwell on generalities further, but rather, continue the discussion with specific propositions about primary processes whose role in the shaping of the EBE state of the metacommunity has to be clarified:

**Proposition 1.** *The phylogenetic process left measurable footprint in the metacommunity's energy structure.*

Since such a footprint is related to floristic and inherent

---

[8] I mention in this connections the mid-20th Century work of the late French-Canadian bio-geographer, Pierre Dansereau, whose long lists of "laws" cover many very specific cases as yet untested.

functional trait diversity, it is analytically recoverable by entropy-based quantum analysis (discussed in the sequel).

**Proposition 2**. *Current environmental mediation modifies the energy structure by forcing compositional change (transience) in the metacommunity.*

The change happens in situ, and could be called facilitation. The effect on the energy structure is analytically recoverable by quantum analysis of temporal or quasi-temporal catena.

**Proposition 3**. *The natural process is never completely deterministic, nor ruled completely by chance.*

A logical consequence of this is the definition that the core objective in quantum analysis is the analytical isolation of the energy structures components issuing from the determinism of phylogeny and environment mediated transience. In practice, any other energy left over is assigned to the grab bag of chance generated events.

These propositions do in deed delineate a multistate, hierarchical quantum analysis of the metacommunity records.

But what is that "energy" whose structure we are discussing?

# What is energy?

I mentioned "energy" already so many times, but for its definition I cannot do better than the Nobel Laureate physicist Richard Feynman[9,10] did in his introductory physics course to undergraduates[11] at Caltech: "... we have no knowledge what energy is", he said. This is not to say that there is no such a thing as energy. All of us are far too well familiar from both pleasing and unpleasing experiences of *energy's manifestations*. As a matter of fact, the measurement of the manifestations is the only way to handle the energy measurement problem.

The manifestations are the energy *footprints*. They are, of course, numerous and very different. So, one should not be surprise to find if the list of energy units used at large takes up 31 pages in Wikipedia[12].

Einstein's equation $E=mc^2$ is my first example - and I wonder if it means anything immediately relevant in the exploration of the energy structure in metacommunities.

A less esoteric case, $E=mgh,$ usually given in "kg meter/s²"

---

[9] http://en.wikipedia.org/wiki/Energy -- Richard Feynman, The Feynman Lectures on Physics (1964) Volume I, 4-1

[10] http://en.wikipedia.org/wiki/The_Feynman_Lectures_on_Physics

[11] Most students in class, according to hearsay, did not really benefit from Feynman's lectures. The experiment of an expert from the highest echelons of science teaching an introductory course had to be reconsidered.

[12] http://en.wikipedia.org/wiki/Category:Units_of_energy

units, measure the *work* needed to lift up an object of mass $m$ against gravitational acceleration $g$ to height $h$. This is fine when the object is a piece of rock. But how much is there in this for the ecologist interested in, for example, the energy structure in metacommunities? "Work" is involved which creates that structure, but not in a simple mechanical act, such as lifting a piece of rock up to some specific height. The "work" is in the complex processes of phylogeny, current environmental mediation, and of course, the chance events. What could be an obvious manifestation of the work done in those cases? I use the plant trait "ground cover" in the examples. I could have used density, biomass, or anything else ecologists use to measure plant or plant community performance.

# *The energy unit*

Regarding the choice of the energy unit, we are in a free-for-all or as-we-please situation. We can chose the energy unit in any way that suits best our well-considered objectives. I opted for "natural units" or "nats". If one thinks about it, it is very quickly resolved that the energy we are interested in is potential energy, and Max Planck's scalar function for this is *energy-based entropy* (EBE ). This is something many of us used to know in its roughest outlines from high school physics, but probably forgot. I rehash some of the technical terms.

In the recapitulation of the original terminology, our first act is to strip the hierarchical relevé of its ecological label and refer to it as a *complex* or more explicitly a *complex of resonators*. We rename the taxa as *resonators*. These terms have functionality on any level of the phylogenetic hierarchy and in any concatenation or stacking of the hierarchical relevés.

On the "0" level of the hierarchy (Figure 3) we have three data vectors which we bring into the discussion as the row vectors:

| | Resonator | | | | | | |
|---|---|---|---|---|---|---|---|
| Complex | 1 | 2 | 3 | 4 | 5 | 6 | Total |
| RL1 | 5 | 7 | 3 | 2 | 2 | 4 | 23 |
| RL2 | 7 | 9 | 9 | 1 | 8 | 8 | 42 |
| RL3 | 4 | 9 | 3 | 9 | 7 | 7 | 39 |
| Total | | | | | | | 104 |

There have cumulant vectors also, one for each hierarchical level:

| Level 1 | Resonator | | | | |
|---|---|---|---|---|---|
| Complex | 1 | 2 | 3 | 4 | Total |
| RL1 | 5 | 10 | 4 | 4 | 23 |
| RL2 | 7 | 16 | 9 | 8 | 42 |
| RL3 | 4 | 12 | 16 | 7 | 39 |
| Total | | | | | 104 |

| Level 2 | Resonator | | |
|---|---|---|---|
| Complex | 1 | 2 | Total |
| RL1 | 15 | 8 | 23 |
| RL2 | 25 | 17 | 42 |
| RL3 | 16 | 23 | 39 |
| Total | | | 104 |

Each number in any of the tables represents an *energy unit count*. For example, we take number 9 and read "nine energy units", 23 "twenty three energy units", and so forth. If we denote the amount of EBE of one unit by the Greek letter epsilon $\varepsilon$, the energy in the complex is $104\varepsilon$ in total, $23\varepsilon$ for RL1, $42\varepsilon$ for RL2 and $39\varepsilon$ for RL3.

# Energy-based entropy

The fundamental quantity is a high-level diversity function:

$$H_n = k \ln W + \text{constant}$$

This is known as Max Planck's *energy-based entropy*. We see in the equation two constants and W. To avoid the symbols clashing with the usual symbols of statistical ecology, I write the same equation with P for W:

$$H_n = -k \ln P + \text{constant}$$

In this $P = \dfrac{1}{C}$ , and further,

$$C = \frac{(n+T-1)!}{(n-1)!\,T!} \approx \frac{(n+T)^{n+T}}{n^n T^T}$$

In this n is the number of resonators in the complex, whose total energy unit count is T. In other words, P is the probability of drawing a complex at random from C distinct complexes whose resonator number is n and total energy unit count is T. The complexes are equiprobable. The last term in the combinatorial equation is based on Stirling's approximation for factorials. This is convenient to use in calculations and algebraic derivations. If proportionality is sufficient, we can leave off the constants and use as a working equation:

$$H_n = nH = (T+n) \ln (T+n) - T \ln T - n \ln n$$

The one resonator portion of $H_n$ is

$$H = \frac{H_n}{n}$$

When we are given an H value, we can get the probability by $P = e^{-H}$.

Where is the energy connection in $H_n$? Planck's 1901 paper makes this absolutely clear, but not so simply. I extract the following from the original paper which I feel shall answer the question by appealing to intuition:

1. Let $U_n = nU$ be the total energy of the complex. Then $U = \dfrac{U_n}{n}$ is the energy of one resonator. Given T, the total number of energy units, and $\varepsilon$ an energy element (one quantum) then $nU = T\varepsilon$.

2. Substitute $\dfrac{nU}{\varepsilon}$ for T in the working equation to obtain:

$$nH = \ln\ C = (n+T)\ \ln\ (n+T) - n\ \ln\ n - T \ln\ T$$

$$= n\left[\left(1 + \frac{U}{\varepsilon}\right)\ln\left(1 + \frac{U}{\varepsilon}\right) - \frac{U}{\varepsilon}\ \ln\ \frac{U}{\varepsilon}\right]$$

Further, divide by n to obtain

$$H = \left(1 + \frac{U}{\varepsilon}\right)\ln\left(1 + \frac{U}{\varepsilon}\right) - \frac{U}{\varepsilon}\ \ln\ \frac{U}{\varepsilon}.$$

What do we see? When we use EBE , the potential energy state of the T totalled n resonator complex is made a function of its uniqueness

$$P = \frac{1}{C}$$

At one point we may consider the value of P so small that the assumption of the complex in hands being a chance event is no longer tenable. At that point we may decide to opt for the alternative assumption that the complex is the product of a deterministic process. When we select this option, two things happen:

1. The realm of statistical decision making enters the analysis, namely, we start distinguishing events that are unique and therefore significant, from those that are not unique. In

this the conceptual difficulties go far beyond the mechanistic examination of the value of P.

2. We find ourselves facing the well-known paradox of making a decision on the basis of a given P, whose justification is in the assumption that the complex is a truly chance generated event, but then we invalidate P by concluding a deterministic origin for the complex.

# Potential energy structure

## The energy equation

For us the energy equation has proxy in the :

$$n_{cx,res}H(cx,res) = n_{cx}H(cx|res) + n_{res}H(res|cx) + n_{cx;res}H(cx;res)$$

The terms in the equation represent the EBE footprints of specific processes. The above quantities identify segments in a Venn diagram (Figure 4) such that $n_{cx}H(cx|res)$, $n_{res}H(res|cx)$, $n_{cx}H(cx)=A+B$ and $n_{res}H(res)=B+C$. Further, $n_{cx;res}H(cx;res)=B$ and $n_{cx.resH}(cx,res)=D$.

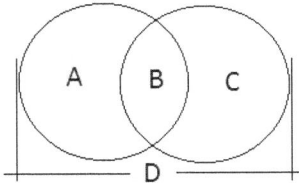

Figure 4. Energy structure's components. Note that A and C are equivocation terms. D is the joint, B the shared (mutual, emergent), and A+C+2B the total EBE .

The energy equation defines a three-parted EBE structure. Symbol n stands for the energy unit counts. The individual terms are manifestation of a joint effect $n_{cx,res}H(cx,res)$, a specific effect $n_{cx}H(cx|res)$ for the catena, a specific effect, and $n_{res}H(res|cx)$ for phylogeny. The emergent effect $n_{cx;res}H(cx;res)$ encapsulates all effects which lack determinism. The terms' changing proportions in time and space tell the story of the metacommunities temporal and spatial energy dynamics in EBE terms.

The energy unit count n depends on the term in the equation. In general,

$n_{cx}$ – number of complexes (hierarchical relevés). This is a fixed number.

$n_{res}$ – number of resonators. This number usually refers to the metacommunity elements (plant taxa), number of nodes on a hiarrachical level, or number of parts (time points, spatial loci, etc.) of the catena.

$n_{cx}H(cx)$ –EBE footprint of the given complex such as the metacommunity.

$n_{cx}H(cx|res)$ - EBE footprint specific to the extrinsic forcing process (say grazing) in environmental mediation of transience.

$n_{res}H(res|cx)$ – EBE footprint specific to the long-term phylogenetic process (mapped by the taxonomic hierarchy, a local proxy for the true phylogenetic tree), or to the temporal process of recovery from perturbation.

$n_{cx;res}H(cx;res)$ - EBE footprint specific to emergent effects.

$n_{cx;res}$ can be determined by interpolation based on T and $n_{cx;res}H(cx;res)$.

$n_{cx,res}H(cx,res)$ – EBE footprint of the joint effect .

$n_{cx,res}=n_{cx}n_{res}$ where there are no zero entries in the table.

nH and H – EBE state of a complex or one resonator.

P – a probability that nH or H is the outcome of a purely random process.

The formal definition of the individual terms is prerequisite for understanding EBE analysis:

1. $H(cx,res)=\dfrac{1}{n_{cx,res}}\ln C_{cx,res}$ ; $n_{cx,res}=n_{cx}n_{res}$ ; $T=\sum\limits_{i=1}^{n_{cx}}T_i=\sum\limits_{i=1}^{n_{cx}}\sum\limits_{j=1}^{n_{res}}X_{ij}$

T is the grand total of the EBE unit counts X. $C_{cx,res}$ is based on $n_{cx,res}$ and T. In the presence of zero elements, $n_{cx,res}$ may be set equal to the number of non-zero elements in the table.

2. $H(cx|res) = \dfrac{1}{n_{cx}}\ln C_{cx|res} = \dfrac{1}{n_{cx}}\ln \dfrac{C_{cx,res}}{C_{res}C_{cx;res}}$

The implicit T value is the grand total. $C_{cx;res}$ is based on $n_{cx;res}$ (see example).

3. $H(res|cx) = \dfrac{1}{n_{res}}\ln C_{res|cx} = \dfrac{1}{n_{res}}\ln \dfrac{C_{cx,res}}{C_{cx}C_{cx;res}}$

Here too, the implicit T is the grand total.

The reason for calling $H(res|cx)$ the EBE footprint of the phylogenetic process per metacommunity element is this: phylogeny is the cause of taxon (resonator) richness in the flora and the arbiter of self-organisation (Orlóci 2012) in the metacommunity. Phylogeny's inertia vector is therefore pointed against mediated transience, forced by extrinsic effects (grazing could be one) on the temporal catena.

4. We expect an emergent effect whose EBE footprint is

$$H(cx;res) = \dfrac{1}{n_{cx;res}}\ln C_{cx;res} = \dfrac{1}{n_{cx;res}}\ln \dfrac{C_{cx,res}}{C_{cx}C_{res}}$$

The implied T is the grand total and $n_{cx;res}$ (see example).

## *Moving the baseline up*

We can define for each level in the hierarchy two types of EBE footprints. One is

$$n_{res,i}H(cx)_i = \ln C_{cx,i} \quad \text{or} \quad H(cx)_i = \dfrac{1}{n_{res,i}}\ln C_{cx,i}$$

In this $n_{cx,i}$ is the number of nodes on hierarchical level i. The implicit T is the grand total of the energy unit counts of the complex so that

$$C_{cx,i} = \frac{(n_{res,i}+T)^{n_{res,i}+T}}{n_{res,i}^{n_{res,i}}T^{T}}$$

The other is a difference,

$$dn_{res,i}H(cx)_i = n_{res,0}H(cx)_0 - n_{res,i}H(cx)_i = \ln \frac{C_{cx,0}}{C_{cx,i}}$$

$$dH(cx)_i = \frac{dn_{res,i}H(cx)_i}{n_{res,0} - n_{res,i}}$$

These tell us how much the per resonator energy-based entropy footprint should shrink if we moved the baseline up from Level 0 to Level i. What is the point of the exercise? We gain flexibility by being able to measure how much loss to be expected in precision if the metacommunity elements are defined at the genus level or higher.

Other more specific cases of the energy equation will be considered as necessary within the examples to be presented. These concern differences in the EBE footprints.

# *Further on P*

Any of the terms in the energy equation can be associated with a P, whose one complement is a relative measure of uniqueness. The P values come from C or H,

$$P = \frac{1}{C} = e^{-H}$$

Statistically speaking, the smaller the value of P the more plausible is the assumption that the complex is the product of a deterministic process and not the outcome of pure chance. It is in fact accepted practice not to assume dominance of the random process in energy dynamics when P is reasonably small, say less than 0.05 or H is at least 3.

# Numerics

## Preliminaries

We now return to the T and n values we have seen in Figure 3, and do some calculations:

| Level | Complex RL1 | | Complex RL2 | | Complex RL0 | |
|---|---|---|---|---|---|---|
| | T | $n_{res}$ | T | $n_{res}$ | T | $n_{res}$ |
| 2 | 23 | 2 | 42 | 2 | 39 | 2 |
| 1 | 23 | 4 | 42 | 4 | 39 | 4 |
| 0 | 23 | 6 | 42 | 6 | 39 | 6 |

The first set of results:

| RL1 | C | $n_{res}H(cx)$ | dC | $dn_{res}H(cx)$ | $dn_{res}$ | $dH(cx)$ | P |
|---|---|---|---|---|---|---|---|
| Level 2 | 1063 | 6.97 | 1063 | 6.97 | 2 | 3.48 | 0.03 |
| Level 1 | 82955 | 11.33 | 78 | 4.36 | 2 | 2.18 | 0.11 |
| Level 0 | 2635689 | 14.78 | 32 | 3.46 | 2 | 1.73 | 0.18 |

| | C | $n_{res}H(cx)$ | $n_{res}$ | $H(cx)$ | P |
|---|---|---|---|---|---|
| Level 0 | 2635689 | 14.78 | 6 | 2.46 | 0.09 |

*Note: numbers are rounded to the decimals to fit the tables on the page.

We take Level 1 and recap the steps:

$$dC_{cx,1} = \frac{82954.78}{1063} \approx 78$$

$$dn_{res,1}H(cx)_1 = \ln dC_{cx,1} = 4.36$$

(the same as 11.33 - 6.97)

$$dn_{res,1} = 4 - 2 = 2$$

$$dH(cx)_1 = 4.36/2 = 2.18$$

$$P = e^{-2.18} = 0.113$$

the same as,

$$P=\frac{1}{dn_{res,1}\sqrt{dC_{cx,1}}}=0.113$$

Note, the calculation of C can be bypassed by going directly to nH:

$$n_{res}H(cx)=(23+4)\ln(23+4)-23\ln23-4\ln4=11.32605$$

Further explanations are in order:

1. When we define a complex as a hierarchical relevé we have to make provision for the upward dependence of the levels in the hierarchy. We measure this by the dH(cx) difference.

2. The $dC_{cx,i}$, $dn_{res,i}H(cx)_i$, $n_{res,i}$ and $dH(cx)_i$ are hierarchical level specific quantities. The inverse of $dC_{cx,i}$ is a probability and $dH(cx)_i$ is the specific contribution of Level i per resonator to the EBE structure of the complex.

3. The probability 1-P is a direct measure of the H value's uniqueness. The smaller is P or the larger is 1-P, the more likely that H is the consequence of a deterministic process.

Similar calculations are performed as before on RL2 and RL3:

| RL2 | C | $n_{res}H(cx)$ | dC | $dn_{res}H(cx)$ | $dn_{res}$ | dH(cx) | P |
|---|---|---|---|---|---|---|---|
| Level 2 | 3415 | 8.14 | 3415 | 8.14 | 2 | 4.07 | 0.02 |
| Level 1 | 798271 | 13.59 | 234 | 5.45 | 2 | 2.73 | 0.07 |
| Level 0 | 71482823 | 18.08 | 90 | 4.49 | 2 | 2.25 | 0.11 |

| | C | $n_{res}H(cx)$ | $n_{res}$ | H(cx) | P |
|---|---|---|---|---|---|
| Level 0 | 71482823 | 18.08 | 6 | 3.01 | 0.05 |

| RL0 | C | $n_{res}H(cx)$ | dC | $dn_{res}H(cx)$ | $dn_{res}$ | dH(cx) | P |
|---|---|---|---|---|---|---|---|
| Level 2 | 2955 | 7.99 | 2955 | 7.99 | 2 | 4.00 | 0.02 |
| Level 1 | 601705 | 13.31 | 294 | 5.32 | 2 | 2.66 | 0.07 |
| Level 0 | 47221083 | 17.67 | 78 | 4.36 | 2 | 2.18 | 0.11 |

| | C | $n_{res}H(cx)$ | $n_{res}$ | H(cx) | P |
|---|---|---|---|---|---|
| Level 0 | 47221083 | 17.67 | 6 | 2.95 | 0.05 |

We can say with certainty based on H(cx) that at least numerically the one resonator energy footprint of RL1 is lowest. Saying this may not be sufficient. We have to make provision in a more complete statement for the statistical sampling error which affects our perception of the statistical uniqueness of H. The value of P comes handy for this. Regarding the interpretation of P, as already mentioned, we assume that the complex in hands is one of C equally probable complexes in the normal spectrum. Therefore it follows that in any test of uniqueness the associated H has to be assumed *a priori* not unique. Whether we continue holding to this assumption after the test, or cast it away as being untenable, will depend on our interpretation of P. For example, upon examination of the differences 3.01-2.46=0.55, 2.95-2.46=0.49 and 3.01-2.95=0.06, we find that the corresponding P values are 0.58, 0.61, and 0.94. Based on these we have to conclude that the differences cannot be considered statistically significant. In other words, on a one-resonator basis the three EBE footprints are not significantly different.

There are different directions the analysis may take at this particular junction.

# *Catenation*

The three RL complexes in the foregoing are taken from three consecutive points on a temporal catena. In the next step we concatenate RL1, RL2 and RL3 into a single hierarchy with 6, 12 and 18 nodes on the Levels 2, 1, 0. We want to test the hypothesis that the catena's EBE footprint per resonator is unique. If we find it such, then we conclude that the temporal catena has left a significant footprint in the EBE structure.

The following are the energy unit counts and the hierarchical level-dependent test criteria in computer output format:

```
Energy unit counts and cumulants by hieratical level
level 2: 15 8 25 17 16 23
level 1: 5 10 4 4 7 18 9 8 4 12 16 7
level 0: 5 7 3 2 2 4 7 9 9 1 8 8 4 9 3 9 7 7

Level totals T and lengths n
 104 6 104 12 104 18

C* Hn H P by level
level 2 :-- 23.28563 3.8809383 2.0631457e-2
level 1 :-- 38.580929 3.2150774 4.0152226e-2
level 0:-- 51.047222 2.8359568 5.8662371e-2

*Computation of C bypassed
```

We can draw up a readable table and examine the numbers:

| | C | $n_{res}H(cx)$ | $dn_{res}H(cx)$ | $dn_{res}$ | $dH(cx)$ | P |
|---|---|---|---|---|---|---|
| Level 2 | -- | 23.29 | 23.29 | 6 | 3.88 | 0.02 |
| Level 1 | -- | 38.58 | 15.30 | 6 | 2.55 | 0.08 |
| Level 0 | -- | 51.05 | 12.47 | 6 | 2.08 | 0.13 |
| | | $n_{res}H(cx)$ | $n_{res}$ | $H(cx)$ | P | |
| Level 0 | -- | 51.05 | 18 | 2.84 | 0.06 | |

The specific $dH(cx)$ value contributed by Level 2 is definitely unique (small P value) and come close to be such on Level 1, definitely not on Level 0. The EBE footprint of the joint effect $H(cx)$ is unique.

Our next task is to determine the amount of reduction in EBE should we move the baseline of recording up to level 1. The reduction is the amount $dH(cx)$ specific to Level 0, which is 2.08 nats per resonator.

The analysis of the concatenated RL records indicated a significant $H(cx)$ for Level zero. We can ask the question whether the RLs contributed differentially to the EBE structure of the concatenated complex on a per resonator basis. We examine the $H(cx)$ values in the table below to find out if in fact they did:

| Level 0 | C | $n_{res}H(cx)$ | $n_{res}$ | $H(cx)$ | P |
|---|---|---|---|---|---|
| RL 1 | 2635689 | 14.78 | 6 | 2.46 | 0.09 |
| RL 2 | 71482823 | 18.08 | 6 | 3.01 | 0.05 |
| RL 3 | 47221083 | 17.67 | 6 | 2.95 | 0.05 |

The dH(cx) values are as follows:

RL 1 to RL 2: 0.55   RL 1 to RL 3: 0.49   RL 2 to RL 3: 0.06

The associated P values (0.58, 0.61, 0.94) indicate statistically insignificant differences on a per resonator basis. A case of footprint size equality is a reasonable assumption.

# Complexes as rectangular arrays

What could be a reason calling for analysis of such an arrangement of the relevé complexes? Typically, when environmental mediation is not on a catena but distributed over the landscape in some random manner, the rectangular array gives a non-catenal (contingency table) mapping of the response levels. The analysis follows what we usually do in canonical contingency table analysis, but at this time it uses totals, and leave untouched the individual cells inside the contingency table.

The entropy unit counts are reproduced below for RL1, RL2 and RL3 in a 3 x 6 table. The following is the computer printout taken as found:

```
3x6 contingency table
 5 7 3 2 2 4
 7 9 9 1 8 8
 4 9 3 9 7 7
Complex (row) totals
 23 42 39
Resonator (column) totals
 16 25 15 12 17 19

Grand total 104
```

The analysis culminates in EBE footprint estimates::

```
(A)  Resonators(columns)
  C* nH H P
  -- 23.28563 3.8809383 2.0631457e-2
(B)  Complexes (relevés, rows)
  C* nH H P
  -- 13.680195 4.560065 1.0461379e-2
(C)  Joint (all causes)
  C* nH H P
  -- 51.047222 2.8359568 5.8662371e-2
(D)  Emergent effects (shared effect)
  C* nH H P
  -- 14.0814 4.6938  0.009
```

**\*Computation of C is bypassed in all cases.**

We construct an annotated table of the results:

| Complexes | $n_{cx}H(cx|res)$ | $n_{cx}$ | $H(cx|res)$ | % | P |
|---|---|---|---|---|---|
| | 13.680195 | 3 | 4.560065 | 26.80 | 0.0105 |
| Resonators | $n_{res}H(res|cx)$ | $n_{res}$ | $H(res|cx)$ | | |
| | 23.28563 | 6 | 3.8809383 | 45.62 | 0.0206 |
| Joint | $n_{cx,res}H(cx,res)$ | $n_{cx,res}$ | $H(cx,res)$ | | |
| | 51.047222 | 18 | 2.8359568 | 100 | 0.0587 |
| Emergence | $n_{cx;res}H(cx;res)$ | $n_{cx;res}$ | $H(cx;res)$ | | |
| | 14.0814 | *3 | 4.6938 | 27.59 | 0.0092 |

\* by iteration.

A further note is in order. It can be shown that with decreasing $|n_{cx} - n_{res}|$, the size of $n_{cx;res}H(cx;res)$ increases and the energy-based entropy structure's stability declines.

The energy equation sections should be consulted for generalised symbols in the equation:

$$n_{cx,res}H(cx,res) = n_{cx}H(cx|res) + n_{res}H(res|cx) + n_{cx;res}H(cx;res)$$

The ranking of the contributions from largest to smallest: $E_{Phy}$, $E_{Env}$, $E_{Rnd}$. Considering that we based the example on an arbitrary small data set, we stop with the further discussion of the results. Now we move on to the analysis of real, field data from two sources.

# Coquihalla case study

## Site

This example uses the M. Mihály data set, a portion of which I present in Table 1. I described the site and sampling design to some length in connection with Figure 2.

The site is inside a meander on the Coquihalla River floodplain in Hope, British Columbia. A common characteristic of floodplains is the presence of more or less level grounds at different elevations Levels are formed by the joint effect of erosion and sedimentation and have a slight incline toward the adjacent higher terrain. The levels are called benches when flooded periodically or terraces when out of reach for floods from the river.

The surveyed transect crosses three well-defined natural levels with elevation increasing as distance from the river increases. The average elevation of the two benches is 4.2 m to 5.4 m above the water level in the river on the first day of the survey. The average height of 3rd level, a terrace, is 10.8 m.

The vegetation cover is remarkably different by appearance on the different levels and uniform on the same level[13].

---

[13] The uniformity and also the differences are indications of different overflow frequencies, overflow duration, and the quality and quantity of the sediment load carried by the flood waters to the sites.

# Data set

M. Mihály's data set give a choice between different data types. I opted for cover/abundance estimates recorded for 73 species in 45 sample plots 10m x 10m each. A subsample of 17 species are shown in Table 1 on which quantum analysis was performed. The chosen species occur on all levels of the floodplain and terrace.

RL1, RL2 and RL3 designate the metacommunities on the three levels. The grand totals and systematic status code (Order, Class, Family, Genus, and Species) are listed in Table 1. Parameter n takes on values depending on the designs of quantum analysis. The definition of the n, nH and H related symbols is the same as in the previous sections.

Table 1. The reduced Coquihalla data set. Brief description of site and sampling design appears in the main text. Plant identification follows the standard field manuals.[14] Table headings: # -- sequence number in M. Mihály's original records; CD – code vectors identifying species mappings in evolutionary plant systematics; a, b, c – floodplain levels low to high. The $C/A$ totals (third segment in the table) are based on 14, 20 and 11 sample plots laid on levels a, b, c.

| # | Plant taxa (species) | Class | CD | Order | CD |
|---|---|---|---|---|---|
| 3 | Acer macrophyllum (shrub) | Eudicots | 3 | Sapindales | 11 |
| 60 | Symphoricarpos albus | Asterids | 1 | Dipsacales | 3 |
| 43 | Polystichum munitum | Pteridopsida | 5 | Denntaedtiales | 2 |
| 37 | Mnium spinulosum | Bryopsida | 2 | Eubrya | 5 |
| 54 | Rubus spectabilis | Magnoliopsida | 4 | Rosales | 10 |
| 65 | Trientalis latifolia | Eudicots | 3 | Ericales | 4 |
| 49 | Rhytidiadelphus loreus | Bryopsida | 2 | Eubrya | 6 |
| 34 | Mahonia nervosa | Magnoliopsida | 4 | Ranunculales | 9 |
| 24 | Eurynchium oreganum | Bryopsida | 2 | Eubrya | 6 |
| 8 | Amelanchier florida | Magnoliopsida | 4 | Rosales | 10 |
| 52 | Rosa gymnocarpa | Magnoliopsida | 4 | Rosales | 10 |
| 29 | Hylocomium splendens | Bryopsida | 2 | Eubrya | 7 |
| 40 | Pachistima myrsinites | Eudicots | 3 | Celasrales | 1 |
| 43 | Pseudotsuga menziesii | Pinopsida | 5 | Pinales | 8 |

[14] Cronquist, C.L.A. , Owenbey, M. and J.W. Thompson. 1955-1959. Vascular plants of the Pacific Northwest. University of Washington Press, Seattle, Washington. Grout, A.J. 1928-1940. Moss flora of North America north of Mexico. Newfane, Vermont.

| # | Species | | | Order | |
|---|---|---|---|---|---|
| 3 | Acer circinatum | Eudicots | 3 | Sapindales | 11 |
| 6 | Achlys triphylla | Magnoliopsida | 4 | Ranunculales | 9 |
| 63 | Thuja plicata | Pinopsida | 5 | Pinales | 8 |
| | Number of states | | 5 | | 11 |

| # | | Family | | Genus | |
|---|---|---|---|---|---|
| 3 | Acer macrophyllum (shrub) | Sapindaceae | 1 | Acer | 1 |
| 60 | Symphoricarpos albus | Caprifoliaceae | 10 | Symphoricarpos | 14 |
| 43 | Polystichum munitum | Dennstaedtiaceae | 7 | Polystichum | 9 |
| 37 | Mnium spinulosum | Mniaceae | 5 | Mnium | 7 |
| 54 | Rubus spectabilis | Rosaceae | 2 | Rubus | 13 |
| 65 | Trientalis latifolia | Myrsinaceae | 4 | Trientalis | 16 |
| 49 | Rhytidiadelphus loreus | Hypnaceae | 6 | Rhytidiadelphus | 11 |
| 34 | Mahonia nervosa | Berberidaceae | 11 | Mahonia | 6 |
| 24 | Eurynchium oreganum | Hypnaceae | 6 | Eurynchium | 4 |
| 8 | Amelanchier florida | Rosaceae | 2 | Amelanchier | 3 |
| 52 | Rosa gymnocarpa | Rosaceae | 2 | Rosa | 12 |
| 29 | Hylocomium splendence | Hypnaceae | 6 | Hylocomium | 5 |
| 40 | Pachistima myrsinites | Celastraceae | 9 | Pachistima | 8 |
| 43 | Pseudotsuga menziezii | Pinaceae | 3 | Pseudotsuga | 10 |
| 3 | Acer circinatum | Sapindaceae | 1 | Acer | 1 |
| 6 | Achlys triphylla | Berberidaceae | 11 | Achlys | 2 |
| 63 | Thuja plicata | Cupressaceae | 8 | Thuja | 15 |
| | Number of states | | 11 | | 16 |

| Species | RL1 | RL2 | RL3 |
|---|---|---|---|
| Acer macrophyllum (shrub) | 23 | 3 | 5 |
| Symphoricarpos albus | 70 | 18 | 4 |
| Polystichum munitum | 105 | 47 | 1 |
| Mnium spinulosum | 69 | 17 | 2 |
| Rubus spectabilis | 8 | 43 | 16 |
| Trientalis latifolia | 7 | 38 | 19 |
| Rhytidiadelphus loreus | 25 | 47 | 34 |
| Mahonia nervosa | 37 | 132 | 43 |
| Eurynchium oreganum | 42 | 99 | 51 |
| Amelanchier florida | 3 | 8 | 16 |
| Rosa gymnocarpa | 2 | 8 | 15 |
| Hylocomium splendence | 23 | 123 | 73 |
| Pachistima myrsinites | 4 | 65 | 72 |
| Pseudotsuga menziezii | 114 | 121 | 88 |
| Acer circinatum | 106 | 105 | 29 |
| Achlys triphylla | 63 | 94 | 17 |
| Thuja plicata | 38 | 122 | 52 |
| Total (T) | 739 | 1090 | 537 |

# Energy-based entropy numerics

Regarding basic parameters we note:

1. T = 739, 1090, 537 for the three metacommunities; T is the

same for each hierarchical level within a given metacommunity.

2. n = 5, 11, 11, 16, 17 in the basic analyses; 15, 33, 33, 48, 51 in the catena or 17, 3, 3x17 in the rectangular (contingency) matrix.

We introduced the analytical technique in the previous section. The new results are summarised in Tables 2, 3 and in Figure 5.

Table 2. Values of dH(cx) (in top five rows) and H(cx) (in baseline row) for the metacommunities and catena. The relevant T and n are specified in the text.

|  | RL1 | RL2 | RL3 | Catena [RL1 RL2 RL 3] | P |
|---|---|---|---|---|---|
| Class | 5.999 | 6.387 | 5.681 | 6.064 | 0.002 |
| Order | 4.561 | 4.946 | 4.246 | 4.625 | 0.010 |
| Family | 0 | 0 | 0 | 0 | -- |
| Genus | 4.026 | 4.409 | 3.676 | 4.090 | 0.017 |
| Species | 3.824 | 4.206 | 3.513 | 3.888 | 0.020 |
| Baseline | 4.783 | 5.168 | 4.468 | 4.848 | 0.008 |

Table 3. Analysis of the species x relevé array.

| Elevation | $n_{cx}$H(cx\|res) | % | H(cx\|res) | P |
|---|---|---|---|---|
|  | 23.013 | 9.31 | 7.671 | 0.000 |
| Phylogeny | $n_{res}$H(res\|cx) | % | H(res\|cx) |  |
|  | 100.969 | 40.84 | 5.939 | 0.003 |
| Joint | $n_{cx,res}$H(cx,res) |  | H(cx,res) |  |
|  | 247.239 | 100.00 | 4.848 | 0.008 |
| Emergence | $n_{cx;res}$H(cx;res) |  | H(cx;res) |  |
|  | 87.473 | 35.38 | 5.545 | 0.004 |

From what we have discussed so far the attentive reader will deduce that a complex with 3 resonators and given T will have a lower EBE state than the complex which has 17 resonators and the same value of T. An unambiguous statement requires access also to the nH and H quantities. What do nH and H tell us about the Coquihalla metacommunity's EBE structure? -

1. The complex's EBE structure is unique if nH (for the complex), H (for one resonator) or dH (for the differences) exceeds 3.00 or if P is less than 0.05. The entries in the body of Table 2 are considered unique. An over-all strong determinism is indicated.

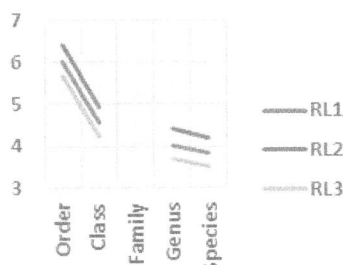

Figure 5. Graphs of dH, based on the first 3 columns in Table 2.

2. The contents of Figure 5 tell us that by moving down through levels in the phylogenetic hierarchy, the EBE footprint of the levels declines on a per taxon basis. The case of Family is left ambiguous by having the same number of nodes as on the Order level.

3. On the Coquihalla transect the elevation incline is sharp. Is there any evidence for differentiation between RL1, RL2 and RL3? Numerically the dH lines are separated on all hierarchical level. But is the separation unique? The greatest difference occurs in the dH values on the "species" level, 4.206 - 3.513 = 0.693. The P value for this is 0.500 which is far too great to any unique dH values. We conclude that the catena is homogeneous, regarding the level-by-level specific per taxa contributions to the EBE structure.

4. We have yet to interpret Table 3. All nH values in the table are unique. Beyond this, we note that in concrete numbers environmental mediation in the manner of the three-step elevation rise comes with an EBE footprint less than $\frac{1}{4}$ as large as the EBE footprint of phylogeny in nH terms. But fortunes

are turned in favour of the elevation gradient $\frac{23}{3}$ to $\frac{101}{17}$ when the H footprints are compared.

The results presented so far are based on the 17 leading species out of about 73 available in the data set. The handling of the data set in its entirety is made difficult by the substantial differences of the levels' metacommunities in species composition. The actual number of species is 46 on Level 1, 54 on Level 2 and 43 on Level 3. This shows that there are many zero entries in the records. The average number of species is about 48 per level, so the average number of zeros is 25.

The grand total of energy unit counts in the large data set is T= 3726. Taking 48 for n, we can do some new calculations:

| | | | |
|---|---|---|---|
| Phylogeny | $n_{res}H(res|cx)$ | $H(res|cx)$ | P |
| | 257.199 | 5.358 | 0.000 |
| Elevation | $n_{cx}H(cx|res)$ | $H(cx|res)$ | P |
| | 24.375 | 8.125 | 0.000 |

We increased the number of species from 17 to 48, and the phylogenetic footprint enlarged. The EBE footprint of environmental mediation stayed more or less the same. This is telling us that even a dramatic change in environmental conditions, such as the level differences on the Coquihalla floodplain, will leave the EBE structure solidly under the control of phylogeny. What can we make of these from an energy point of view? We can retest Pillar's community assembly rule (Pillar and Duarte 2010) in EBE terms:

1. The role of environmental mediation is to facilitate transience by sorting species into metacommunities according to the inherited environmental traits. Therefore, the footprint of environmental mediation in the metacommunity's EBE structure is a function of the environmental breadth n, such

as the number of floodplain levels, over which the T energy units are dispersed. It follows that we can regard the meta-communities as distinct energy entities.

2. The EBE footprint of phylogeny is the consequence of the metacommunity's richness n in resonator equivalent taxa and performance T.

# Campos Sulinos case study

## Site and relevés

I gave a brief introduction to the Campos Sulinos Biome earlier in the text in connection with Figure 1 which shows a picture from the highland Campos. I am using a triplet of relevés from Prof. Dr. Valério Patta Pillar's Campos research files to which he graciously granted access to me. Dr. Carolina Blanco, a principal research associate on Pillar's team, selected the relevés and kindly prepared the basic spreadsheet for quantum analysis. The research site is on the hilly landscape of Eldorado, marked by A on the Google map below:

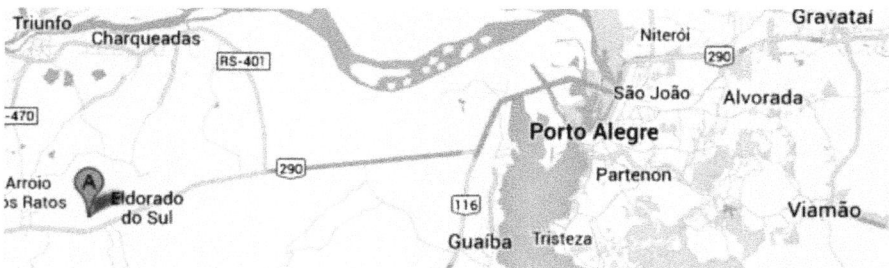

The relevé triplet fixes the momentary states in a meta-community at three successive points in time, prior and posterior to controlled grazing. I label by "0" the relevé of the null state (prior to grazing) and by "47" or "180" the relevé taken 47 or 180 days after grazing. I trimmed the species number back to 50 by discarding the species with zero entries in any of the relevés. The species remaining in the analysis span the entire recording period. Following earlier practice (Pillar and Duarte 2010, Orlóci 2011, 2012, 2013), and with the same justification, I use phylogenetic criteria to create the hierarchical structure in the relevés. The hierarchical levels correspond to Order (Level 4), Class (Level 3), Family (Level 2), Genus (level 1) and Species (Level 0).

The Pillar team performed path analysis on the original multi-trait data set which revealed the fundamental link of multi-trait functional redundancy to the metacommunity's diversity, stability and resilience. What I plan to do in quantum analysis is to bring forth clarifications of the specific effects of grazing and phylogeny on the metacommunity's EBE structure.

## *Quantum analysis*

Propositions stated about the individual terms of the energy equation are tested:

*P1. Two principal processes shape the EBE footprint of metacommunities on grazed Campos. One of these is the long-term phylogenetic process whose vector of inertia points against change. The other is current grazing which forces transience. Which of the two is dominant in the Pillar research site?*

We narrow the analytical problem to isolation of the EBE footprints, one specific to phylogeny and the other to environmental mediation. The basic data arrangement is an $n_{cx}$ by $n_{res}$ contingency table. The EBE function in its task related forms were defined in earlier sections. The results are summarized in Table 4.

Table 4. Level refers to evolutionary systematic status: 4 – Orders, 3 – Class, 2 – Family, 1 – Genus, 0 - Species. Symbols "0", "47" and "180" designate time points at which the relevés were taken before grazing and during the recovery process. Symbols: T⁻ - total number of energy units of the corresponding complex, $n_{res}$ - number of resonators, $n_{cx}$ - number complexes which changes depending on the case, nH - EBE per complex in nats, H - EBE per resonator in nats, P - probability. Guide to some of the calculations:

$n_{res}H(cx) = (645+3) \ln (645+3) - 645 \ln 645 - 3 \ln 3 = 13.555$

$n_{res,cx}H(cx,res) = (1690+6) \ln (1690+6) -1690 \ln 1690 - 6 \ln 6 = 39.819$

$n_{res}H(res|cx) = (1690+2) \ln (1690+2) - 1690 \ln 1690 - 2 \ln 2 = 15.480$

$n_{cx}H(cx|res) = (1960+3) \ln (1960+3) - 1960 \ln 1960 - 3 \ln 3 = 22.004$

$n_{cx;res}H(cx;res) = 39.855 - 22.004 - 15.480 = 2.371$

Part A

| Level | Complex | T | $n_{res}$ | $n_{res}H(cx)$ | H(cx) | P |
|---|---|---|---|---|---|---|
| 4 | "0" | 645 | 2 | 13.555 | 6.7776 | 0.0011 |
| | "47" | 493 | 2 | 13.019 | 6.5093 | 0.0015 |
| | "180" | 552 | 2 | 13.244 | 6.6222 | 0.0013 |
| | Joint | | | $n_{cx,res}H(cx,res)$ | H(cx,res) | |
| | | 1690 | 6 | 39.819 | 6.6365 | 0.0013 |
| | Specific | | | $n_{res}H(res|cx)$ | H(res|cx) | |
| | | 1690 | 2 | 15.48 | 7.7400 | 0.0004 |
| | | | | $n_{res}H(cx)$ | | |
| 3 | "0" | 645 | 9 | 47.511 | 5.2789 | 0.0051 |
| | "47" | 493 | 9 | 45.111 | 5.0123 | 0.0067 |
| | "180" | 552 | 9 | 46.12 | 5.1244 | 0.0059 |
| | Joint | | | $n_{cx,res}H(cx,res)$ | H(cx,res) | |
| | | 1690 | 27 | 138.742 | 5.1386 | 0.0059 |
| | Specific | | | $n_{res}H(res|cx)$ | H(res|cx) | |
| | | 1690 | 9 | 56.141 | 6.2379 | 0.0020 |
| | | | | $n_{res}H(cx)$ | | |
| 2 | "0" | 645 | 13 | 63.886 | 4.9143 | 0.0073 |
| | "47" | 493 | 13 | 60.432 | 4.6486 | 0.0096 |
| | "180" | 552 | 13 | 61.884 | 4.7602 | 0.0086 |
| | joint | | | $n_{cx,res}H(cx,res)$ | H(cx,res) | |
| | | 1690 | 39 | 186.202 | 4.7744 | 0.0084 |
| | Specific | | | $n_{res}H(res|cx)$ | H(res|cx) | |
| | | 1690 | 13 | 76.328 | 5.8714 | 0.0028 |
| | | | | $n_{res}H(cx)$ | | |
| 1 | "0" | 645 | 38 | 146.701 | 3.8605 | 0.0211 |
| | "47" | 493 | 38 | 136.819 | 3.6005 | 0.0273 |
| | "180" | 552 | 38 | 140.966 | 3.7096 | 0.0245 |
| | Joint | | | $n_{cx,res}H(cx,res)$ | H(cx,res) | |
| | | 1690 | 114 | 424.486 | 3.7236 | 0.0241 |
| | Specific | | | $n_{res}H(res|cx)$ | H(res|cx) | |
| | | 1690 | 38 | 182.63 | 4.8061 | 0.0082 |
| | | | | $n_{res}H(cx)$ | | |
| 0 | "0" | 645 | 50 | 179.751 | 3.5950 | 0.0275 |
| | "47" | 493 | 50 | 166.878 | 3.3376 | 0.0355 |
| | "180" | 552 | 50 | 172.275 | 3.4455 | 0.0319 |
| | Joint | | | $n_{cx,res}$ | H(cx,res) | |

| | 1690 | 150 | 518.905 | 3.4594 | 0.0314 |
|---|---|---|---|---|---|
| Specific | | | $n_{res}H(res|cx)$ | $H(res|cx)$ | |
| | 1690 | 50 | 226.755 | 4.5351 | 0.0107 |

## Part B

| Level | $n_{cx,res}H(cx,res)$ | $H(cx,res)$ | P | $n_{cx}H(cx|res)$ | $H(cx|res)$ | P |
|---|---|---|---|---|---|---|
| 4 | 39.85 | 6.64 | 0.001 | 22.00 | **7.33** | 0.001 |
| 3 | 138.90 | 5.14 | 0.006 | 22.00 | **7.33** | 0.001 |
| 2 | 186.43 | 4.78 | 0.008 | 22.00 | **7.33** | 0.001 |
| 1 | 425.14 | 3.73 | 0.024 | 22.00 | **7.33** | 0.001 |
| 0 | 519.75 | 3.46 | 0.031 | 22.00 | **7.33** | 0.001 |

| Level | $n_{res}H(res|cx)$ | $H(res|cx)$ | P | $n_{cx,res}H(cx;res)$ | $H(cx;res)**$ | P |
|---|---|---|---|---|---|---|
| 4 | 15.48 | **7.74** | 0.000 | 2.37 | 2.37 | 0.093 |
| 3 | 56.14 | **6.24** | 0.002 | 60.76 | 4.05 | 0.002 |
| 2 | 76.33 | **5.87** | 0.003 | 88.10 | 3.83 | 0.004 |
| 1 | 182.63 | **4.81** | 0.008 | 220.50 | 3.02 | 0.011 |
| 0 | 226.76 | **4.54** | 0.011 | 270.99 | 2.79 | 0.014 |

\* Numbers in the table are truncated to make the table fit on the page.
\*\* Calculation of the H(cx;res) quantity:

| $n_{cx}H(cx|res)$ | $n_{cx}$ | $n_{res}H(res|cx)$ | $n_{res}$ | $n_{cx,res}H(cx,res)$ |
|---|---|---|---|---|
| 22 | 3 | 15.48 | 2 | 39.85 |
| 22 | 3 | 56.14 | 9 | 138.90 |
| 22 | 3 | 76.33 | 13 | 186.43 |
| 22 | 3 | 182.63 | 38 | 425.14 |
| 22 | 3 | 226.76 | 50 | 519.75 |

| $n_{cx,res}$ | $n_{cx:rc}H(cx;res)$ | $n_{cx;res}$ | $H(cx;res)$ |
|---|---|---|---|
| 6 | 2.37 | 1 | 2.37 |
| 27 | 60.76 | 15 | 4.05 |
| 39 | 88.1 | 23 | 3.83 |
| 114 | 220.51 | 73 | 3.02 |
| 150 | 270.99 | 97 | 2.79 |

Note: recalculated results may not come out an exact match to the table values owing to the rounding errors.

The continuation of Level 4 of Part B is copied below:

| Level | $n_{cx}H(cx|res)$ | $H(cx|res)$ | P | $n_{res}H(res|cx)$ | $H(res|cx)$ | P |
|---|---|---|---|---|---|---|
| 4 | 22.004 | 7.33 | 0.001 | 15.480 | 7.74 | 0.000 |
| | $n_{cx,res}H(cx,res)$ | $H(cx,res)$ | P | $n_{cx;res}H(cx;res)$ | $H(cx;res)$ | P |
| | 39.855 | 6.64 | 0.001 | 2.371 | 2.371 | 0.093 |

What do the numbers tell us? Consider Level 4 in Table 4, Part A (copied here for convenient viewing):

| Level | Complex | T | $n_{res}$ | $n_{res}H(cx)$ | $H(cx)$ | P |
|-------|---------|-----|-----|-----------------|----------|--------|
| 4 | "0" | 645 | 2 | 13.555 | 6.7776 | 0.0011 |
| | "47" | 493 | 2 | 13.019 | 6.5093 | 0.0015 |
| | "180" | 552 | 2 | 13.244 | 6.6222 | 0.0013 |
| | Joint | | | $n_{cx,res}H(cx,res)$ | $H(cx,res)$ | |
| | | 1690 | 6 | 39.819 | 6.6365 | 0.0013 |
| | Specific | | | $n_{res}H(res\|cx)$ | $H(res\|cx)$ | |
| | | 1690 | 2 | 15.48 | 7.7400 | 0.0004 |

It is seen that we deal with additive sequences in the manner of the Venn diagram:

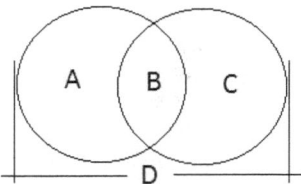

The Venn graph comes from Figure 4. We have for level 4 in Table 4, Part B:

$A=n_{cx}H(cx|res)=22.004$, $B=n_{cx;res}H(cx;res)=2.371$,

$C=n_{res}H(res|cx)=15.45$, $E_{cx}=A+B$, $E_{res}=B+C$.

**Note: A and C are equivocation terms, B is the shared term, and D is the joint term. The total is A+2B+C = D+B.**

Figure 6. Graphs show H values as a function of hierarchical level from Order (Level 4) to Species (Level 0). Symbols: H(cx,res) – joint, H(cx|res) – specific for grazing, H(res,cs) – specific for phylogeny, H(cx;res) – shared or redundant average H. See explanations in the main text.

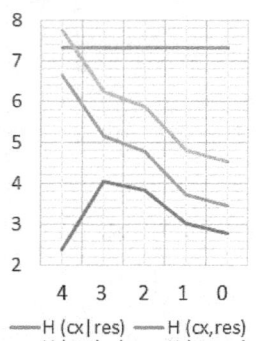

Having completed the arithmetic part, we are now in a position to address the question posed under P1. We focus on the numbers under column headings H(cx|res) and H(res|cx). Three things should become clear on first inspection:

a. The H footprint of grazing is unaffected by phylogenetic level (horizontal line in Figure 6).

b. The H footprint of phylogeny is dependent on the phylogenetic level (second top line in Figure 6).

c. In H footprint of the grazing effect is dominant on all hierarchical levels, except Level 4 where H(res|cx) graph is highest.

d. In nH terms, the dominance of phylogeny is absolute. The discussion at the end of the Coquihalla section applies in the Campos Sulinos case as well.

*P2: The metacommunity, T-totalled and n-valued, is the outcome of an assembly process ruled by chance?*

Table 4 contains the probabilities of interest in the column headed by P. How do we interpret them? For example, the probability that a resonator randomly chosen from all possible resonators with T = 645, n = 2 and H = 6.7776 nats will have an exact match of the energy counts in arrangement as in Complex "0" on Level 4 is 0.0011. What do we conclude about the complex being a random event? Clearly, this is very unlikely. We do better to consider that the complex is the product of the vegetation's response to deterministic effects.

*P3. Change of hierarchical level do not bring about significant change in the value of H?*

We now arrived at the point in the analysis when we decide how significant is the change (dH) in one resonator's energy state when we pass from a higher to a lower taxonomic level. Table 5 and Figure 7 have the relevant results.

Table 5. The first 5 columns match those in Table 4. The others contain the specific nH components, differences between levels in dnH terms, % values relative to 518.9046 nats, the one resonator difference dH=dnH/dn, and the probabilities of dH under the rule of chance. See graphs in Figure 7 and interpretation in the main text.

| Level | Complex | T | $n_{res}$ | $n_{res}H(cx)$ | $dn_{res}H(cx)$ | % | dH(cx) | P |
|-------|---------|------|-----|----------|----------|-------|--------|--------|
| 4 | "0" | 645 | 2 | 13.5553 | | | | |
| | "47" | 493 | 2 | 13.0187 | | | | |
| | "180" | 552 | 2 | 13.2444 | | | | |
| | | | | 39.8184 | 39.8184 | 7.67 | 6.6364 | 0.0013 |
| 3 | "0" | 645 | 9 | 47.5107 | | | | |
| | "47" | 493 | 9 | 45.1112 | | | | |
| | "180" | 552 | 9 | 46.1198 | | | | |
| | | | | 138.7417 | 98.9233 | 19.06 | 4.7106 | 0.0090 |
| 2 | "0" | 645 | 13 | 63.8860 | | | | |
| | "47" | 493 | 13 | 60.4321 | | | | |
| | "180" | 552 | 13 | 61.8836 | | | | |
| | | | | 186.2017 | 47.4600 | 9.15 | 3.9550 | 0.0192 |
| 1 | "0" | 645 | 38 | 146.7012 | | | | |
| | "47" | 493 | 38 | 136.8193 | | | | |
| | "180" | 552 | 38 | 140.9655 | | | | |
| | | | | 424.4860 | 238.2843 | 45.92 | 3.1771 | 0.0417 |
| 0 | "0" | 645 | 50 | 179.7511 | | | | |
| | "47" | 493 | 50 | 166.8782 | | | | |
| | "180" | 552 | 50 | 172.2753 | | | | |
| | | 1690 | 150 | 518.9046 | 94.4186 | 18.20 | 2.6227 | 0.0726 |

We have further points of interest to present regarding Table 5 and Figure 7:

Figure 7. Graph of the recovery parameter dH (vertical scale) across all levels (4 to 0).

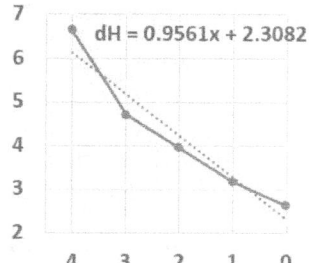

$dH = 0.9561x + 2.3082$

1. Take a block of numbers, any block, say Family (Level 2). This level accounts for 186.2017 nats of nH = 518.9046 nats.

2. Should we decide to move to the next level down to Genera, nH would rise to 424.4860 nats. Corresponding to these we have dnH=238.2843 nats, dn=75, dH=3.1771 nats, and P=0.0471. This P is statistically just reaches significance. In all cases the dH per resonator is positive and is significant with descending magnitude from Order to Species. Is the change in dH significant? At the broadest we have 6.6364-2.6227=4.0137, P =0.018. This indicates a significant change.

None of the other differences are significant. The cut off for significance at 0.05 probability or dH = 3. We can say then that the per resonator EBE footprint depends on the hierarchical level, but the rate of change in dH is steady n a slope of about $45^O$.

*P4. The difference in the H footprint of the complex duplets as actually observed is a consequence of pure chance. In other words, the H footprints have equal expectation.*

Some relevant results are given in the preceding tables. Specific paired comparisons are listed in Table 6.

Table 6. Comparison of complex duplets different hierarchical levels. The symbols are the same as in the forgoing tables. The method and the mode of interpretation is explained in the text.

| Level | Complex duplet | $n_{res}$ | $n_{res}H(cx)$ | H(cx) | dH(cx) | P |
|-------|----------------|-----------|----------------|-------|--------|---|
| 4 | "0" to "47" | 2 | 13.5553 | 0.5366 | 0.2683 | 0.76468 |
|   | "0" to "180" | 2 | 13.0187 | 0.3109 | 0.1555 | 0.85603 |
|   | "47" to "180" | 2 | 13.2444 | 0.2257 | 0.1129 | 0.89328 |
| 3 | "0" to "47" | 9 | 47.5107 | 2.3995 | 1.1998 | 0.30127 |
|   | "0" to "180" | 9 | 45.1112 | 1.3909 | 0.6955 | 0.49885 |
|   | "47" to "180" | 9 | 46.1198 | 1.0086 | 0.5043 | 0.60393 |
| 2 | "0" to "47" | 13 | 63.8860 | 3.4539 | 1.7270 | 0.17783 |
|   | "0" to "180" | 13 | 60.4321 | 2.0024 | 1.0012 | 0.36744 |
|   | "47" to "180" | 13 | 61.8836 | 1.4515 | 0.7258 | 0.48396 |
| 1 | "0" to "47" | 38 | 146.7012 | 9.8819 | 4.9410 | 0.00715 |
|   | "0" to "180" | 38 | 136.8193 | 5.7357 | 2.8679 | 0.05682 |
|   | "47" to "180" | 38 | 140.9655 | 4.1462 | 2.0731 | 0.1258 |
| 0 | "0" to "47" | 50 | 179.7511 | 12.8729 | 6.4365 | 0.0016 |
|   | "0" to "180" | 50 | 166.8782 | 7.4758 | 3.7379 | 0.0238 |
|   | "47" to "180" | 50 | 172.2753 | 5.3971 | 2.6986 | 0.0673 |

How to calculate interpret the results? An example:

1. Take the complex duplet "0" to "47" on Level 4 (Table 6). Observe that for given T=645 and n=2 we have $n_{res}H(cx)$=13.5553 nats. Further, dnH =| 13.0187-13.5553 |=

0.5366 nats, and for one resonator dH = 0.2683 nats. The corresponding probability P is 0.76468.

2. P is telling us that the difference dH is too small to regard it unique.

The dH values help us to put into perspective in EBE terms the effect of elapsed time on recovery:

1. In general, our answer to P4 is that the proposition holds on the higher levels but not on the lower ones.

2. Observe the decline of dH values (Table 6) as recovery time passes from "0" to "47" to "180" on each level. The recovery of the original EBE footprint appears to close the gap rather rapidly. Consider next dH = 4.9410 nats on the Genus level (Level 1, Table 6). This is proportional to the amount of EBE remaining to be regained on a one resonator basis after 47 days in recovery. The nominal amount to be regained on this level of the hierarchy after 180 days shrinks to 2.0731 nats. But this is not significant. We conclude that on the Genus level the EBE footprint recovery process is completed in 180 days.

4. Consider dH = 6.4365 nats on the Species level (Level 0, Table 6). This is proportional to the amount of potential energy left to be regained on a one resonator basis after 47 days in recovery. The nominal amount to be regained after 180 days shrinks to 2.6986 nats. If we consider this insignificant then because P exceeds 0.05, then we have to conclude that the original EBE state is recovered here too in 180 day.

5. Consider the deceleration difference 2.6986-2.0731 = 0.6255 on a per resonator basis between hierarchical levels 0 (Species) and 1 (Genus) in Table 6. Now observe $P=e^{-0.6255}=0.5350$. This is the probability of a differential de-

celeration as large as 0.6255, occurring between the two taxonomic levels under the rule of chance. Similar calculations may be performed for other levels.

# Conjecture driven analysis

The examples make it obvious how EBE analysis works and in what way enable the user to lay open the potential energy structure of metacommunities by proxy. It should be obvious just as much by now that quantum analysis is not another type of diversity analysis, but adds new dimensions to entropy-based analyses. In this regard I refer to Orlóci (2015).

We recall that the record unit, which is allowed in an EBE analysis, is an energy unit count per resonator (species, other taxon, temporal or spatial locus). Important, the analytical penetration still remains on the resonator complex (metacommunity, catena, sample) level. To see this, we need only to consider the fact that the complex's total energy unit count (T) and the number of resonators (n) are directly involved in the parameterisation of E. And further that the probability calculus only takes into account C and leaves the resonator probabilities (p) untouched. This sharply contrast with what is happening when we use the metacommunity's

well-known generalised diversity equation $H_\alpha = \dfrac{1}{1-\alpha} \sum_{i=1}^{n_{res}} p_i^\alpha$

(Rényi 1961) to which so much of diversity studies are linked in Ecology. The scale factor $\alpha$ in Rényi's equation can be 0 or any positive number, except exactly 1.

It is logical to let any analysis to follow a series of well stated conjectures. The Campos analysis contains an éclat example of this:

1. We put the first question about the relative importance of phylogeny and grazing on the potential energy structure of

the metacommunity, the dynamics of which gained momentum by current grazing. The conclusion, after considerable computational labor, projected onto the one-resonator level, is quite remarkable. We find that the grazing effect on the EBE structure is independent from the level of evolutionary plant systematics, but the phylogenetic effect changes. It is strongest for the "Order" level and weakest, yet still significant, for the "Species" level. Redundancy is considerable. This is telling us that strong perturbation raises effects to which the ecologists refer by the term "emergent".

2. The question whether the complexes under scrutiny were in fact not the product of the coincidence of freak chance driven events in the normal spectrum is important. The reason is that any such result would render superfluous any further analysis along the lines as described. The "chance ruled process" hypothesis have been tested. The test indicated rejection of this hypothesis in favour of the alternative, a strong determinism ruled process.

3. A subsequent question required the determination of how much should be the change in H if we take taxa from different levels of phylogenetic hierarchy. We found that the genus level could be taken as the baseline of the systematic hierarchy in at least one case without much penalty by loss of accuracy.

4. Another important point emerged in the analytical process which is addressing recovery velocity in the potential energy structure from grazing. It is obvious from what has been presented that the recovery process is all but complete by the 47th days after grazing stopped to regain the original "0" level EBE structure on higher systematic levels and possibly, but not so definitely on the Species level as well.

# Contrasting models

I had this section up front in the text but I moved it back. It can be read as an introduction given *a posteriori* when the reader gained familiarity with EBE analysis, and especially with my terminology.

Either of the sites which I introduced early on in the text to anchor the discussions into the practitioner's reality (Figures 1, 2) is sufficiently known from ecological studies to allow energy studies to proceed.

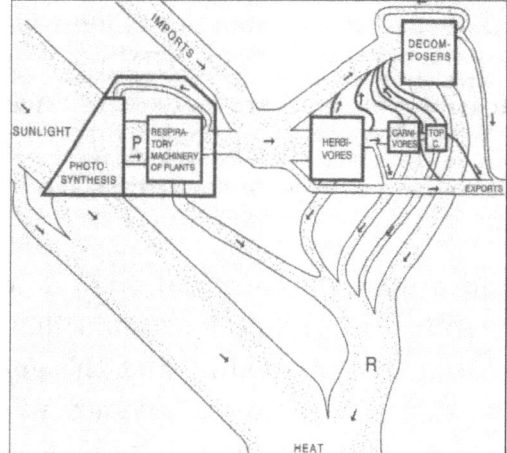

Figure 8. An adaptation of Odum's (1971) model of calorific (trophic) flow (*cf.* S. Maud).[15]

If we followed the conventional way, the study would have the trophic flow of the ecosystem in focus. Our central analytical task would amount to parameterization of some channel-specific trophic equation, perhaps under guidance of a flow chart similar to the one presented in Figure 8. I have considered such a choice and ditched the idea. My reason is the precarious nature of the exercise, knowing full well that I could not possibly isolate in such terms the specific footprints of the

---

[15] This work has been released into the public domain by its author, Sholto Maud at the Wikipedia project. This applies worldwide. http://upload.wikimedia.org/wikipedia/commons/d/d9/Silver_Spring_Model.jpg

fundamental processes in the energy structure.

The conceptualizations of the energy structure as calorific flow vs. EBE come furbished with completely different data analytical techniques. The calorific flow models are well worked and broadly known to students. The conceptual tools which allow us to subject the EBE structure of the meta-community to scrutiny is my first interest in this essay. The first phase in the development of these is the subject of an earlier essay (Orlóci 2013). Further posterior phases include several monographs referenced in Orlóci (2015). This version of the manuscript should be considered the basis of the electronic 2nd edition.

Why should users embrace hitherto insufficiently tested tools and leave on the wayside what ecologists routinely use since what appears to me from ecological science's eternity? I can suggest a few things that led me to the heretic act:

1. I wanted to approach holistically the study of the meta-community's energetics. Why holistically? This is of course the same as asking with Gleick (1987) and Çambel (1993), paraphrased - "Why sciences are turning to holism after investing so much time and resources into reductionism?"

To me the answer is simple: reductionism did not lead science to understand complex systems, like a metacommunity. Therefore, simply stated I approach the analysis of the energy structure by abandoning the idea of calorific flow for a moment, and focussing on the generically holistic property, potential energy. This could be done only by translation of Max Planck's (1901) quantum theoretical principles into workable tools of phytosociology.

2. The EBE approach is not just conceptually well founded, as

already discussed, but it is intensely practical. The EBE approach does in fact open up a vast array of ordinary phytosociological data for use to parameterise the energy equation.

3. The holistic EBE analysis successfully isolated the nH footprints of phylogeny, environmental mediation, and the chance events, for which I developed techniques in moment and product moment terms earlier (see Orlóci 2012).

In closing the main text I should mention three recent analytical models in the manner of statistical and other numerical procedures by which propositions were tested about the assembly rules in metacommunities. They share objectives but differ in their approach to solving the analytical problem of the isolation of signals issuing from the fundamental processes (phylogeny, current environmental mediation, and chance):

1. *The model of Pillar and Duarte* (2010). The algorithm applies the partial correlation as its principle scalar in signal isolation. Through the isolates, Pillar and Duarte revealed that the phylogenetic signal is at play in community assembly when species are sorted among metacommunities according to their inherited environmental traits. This is an important generalization which puts into perspective the arguments in favour of chance playing the dominant role in the assembly of the metacommunity.

2. *My model* (2012). This model similar to the previous one is signal theoretical. It is novel in its use of multiscale hierarchical solutions for the isolation of signals and quantification of their role in the community assembly process. In the case study for testing, the solution have brought to the sur-

face indications of the high stability of functional trait composition and low inclination of the metacommunity for environmentally mediated transience.

3. *My first EBE model* (2013). In this, I incorporated many common aspects of my signal theoretical model, but changed the model parameter to Max Planck's quantum theoretical, energy-based entropy function. My current analytical approach remains focussed on the same objectives.

The models in all cases are algorithms of statistical procedures of some complexity. In the present case the procedure is built around a very simple energy equation of additive terms, each representing the EBE function. The terms are specific to the principal processes' EBE footprints.

I already discussed that the theory which makes the energy function applicable in vegetation studies is Planck's, inwhich energy is conceived not as a continuous wave, but a stream of discrete countable units, the quanta. The "countable energy units" provision enables me to translate Planck's energy-based entropy into operational tools in my field. When I measure a functional trait of the community elements - say ground cover, density, or others used in ecology –, and provided that the adapted Planck's principles are in fact what I think they are, I am actually counting the number of energy units which had to be put to work in the natural process to attain the observed energy state of the trait. The question "How is that work done" is legitimate. But I leave it unanswered in this essay.

By necessity and to the advantage of the exercise, my approach probes ordinary survey or experimental data for process identities and their specific EBE footprints on the metacommunity level. I see the approach to be welcomed by practitioners of the field for two additional reasons. One is the consistent, statistical framework provided for the comparative holistic study of EBE structures and structural dynamics. The second is practicality which should interest practitioners intended to use phytosociological data already in their possession from the past, for new energy studies of the vegetation in any of the major biomes.

# Questions and answers

The interpretation of a computed energy-based entropy value should begin with the recollection that T and n are the quantities used in the computation of C. This implies dependence of the outcome H upon the taxon (resonator) richness (n) of the metacommunity and upon the relevé (complex) total T of the taxa. With all said, I try to anticipate questions regarding quantum analysis:

*1. Why did we have to assume that energy comes in discrete units?*

The proposition that energy comes in discrete, countable units has been a main point in Max Planck's 2001 seminal paper. The assumption is needed for valid application of the equation, $nH = k \ln C + \text{constant}$. Exactly this provision of Planck's quantum theory facilitates the adaption of EBE analysis in vegetation studies.

*2. What justifies calling nH energy-based entropy?*

This is another point Max Planck makes in his 2001 paper. He gives as the proof in the manner of the equation,

$$nH = k \ln C = k \ln \frac{(n+T)^{n+T}}{n^n T^T} = kn\left[\left(1+\frac{U}{\varepsilon}\right)\ln\left(1+\frac{U}{\varepsilon}\right) - \frac{U}{\varepsilon}\ln\frac{U}{\varepsilon}\right]$$

*3. Where is the probabilistic connection of nH?*

When we assume discrete, countable energy units, and assume further the rule of chance over the assembly of an observed complex (hierarchical relevé, metacommunity) with a given n and a given T, the complex we actually have is one of

$C = \dfrac{(n+T)^{n+T}}{n^n T^T}$ equiprobable complexes. Each has the same

chance of materialising in a chance ruled Normal process. Therefore the hierarchical relevé, as we observed it, has a probability of assembling exactly the same way as it actually

is by chance alone, equal to $\dfrac{1}{C}$. When this has a small value

(on the 0 to 1 scale), the relevé of the metacommunity is considered unique. When it has a large value, the relevé is considered common.

*4. We accepted from Max Planck that nH is a proxy parameter of the complex's energy level.*

What kind of energy? The energy implied is potential energy.

*5. Why do we use the phrase "contribution to the energy state".*

Energy is not measurable directly, only by its manifestations, the energy footprint. We use "footprint" because it sounds right that way - as a metaphor. Our unit of measure is the *nat*, because the measuring function we elected to use is $E = \ln P$.

*6. Why do we use hierarchical relevés?*

The hierarchical relevé puts a handle on the problem of interconnecting the principle processes which account for directed variation in n and T. In our examples, one is phylogeny and the other current environmental mediation.

*7. Are taxa and environmental variables not left out of the conclusions when we apply the methodology just described?*

True, the conclusions we draw are at least one step removed from the community elements and also from the environmental variables. Therefore, any comparison made between complexes (hierarchical relevés or levels in hierarchical relevés) is a comparison of their potential energy. In this regard the methodology joins the group of information theoretical techniques whose purpose is the interrogation of the community's diversity structure. By so doing, the discovery of any governance rules is strictly in such terms.

*8. What is the ultimate statement we can make after completing the EBE analysis?*

In the briefest, it is a probabilistic statement, regarding the potential energy structure of the metacommunity, a condition of phylogeny, environmental mediation and emergent effects. An important part of this is the interpretation of the probability P. Take the example $T = 645$, $n = 2$, $H = 6.7776$ nats and $P = 0.0011$. The correct statement about P goes like this: the probability that a resonator complex randomly chosen from all possible resonator complexes with $T = 645$, $n = 2$ and $H = 6.7776$ nats will have an exact match of energy counts as arranged in the metacommunity's relevé.

*9. Finally, to the reader who lost track of the meaning of "complex" and "resonator", or think that the terms "energy" and "energy unit" are confusing, I reemphasize these points:*

a. Think of a complex as an object described by the states of n resonators. If the "complex" designates a plant community, its "resonators" are usually the community elements, in some way defined as taxa.

b. Think of a complex as an entire hierarchical relevé. The resonators are still taxa, but the energy structure is hierarchical.

c. Do not forget that the analysis is based on potential energy unit counts. Furthermore, the parameter is EBE and this is dual to potential energy.

# Reference bibliography

Çambel A.B. 1993. Applied Chaos Theory: a Paradigm for Complexity. Academic Press, New York.

Camazine, S., Deneubourg, J.L., Franks, N.R. Sneyd, J., Theraulaz G. and E. Bonabeau. 2003. Self-Organization in Biological Systems, Princeton University Press.

Dansereau, P. 1954. Biogeography. An ecological perspective. Ronald Press, New York.

Diaz, S., Acosta, A., and M. Cabido. 1994. Grazing and the phenology of flowering and fruiting in a montane grassland in Argentina – a niche approach. Oikos 70: 287-295.

Felsenstein, J. 2004. Inferring Phylogenies. Sinauer Associates, Sunderland, MA.

Gleick, J. 1987. Chaos. Making a New Science. Penguin Books, New York.

Henning, W. 1965. "Phylogenetic Systematics," Ann. Rev. Entomol., Vol. 10, 97-116

Huxley, J. 1942. Evolution: the modern synthesis. The MIT Press.

Kerner von Marilaun, A. 1863. Das Pflanzenleben der Donauländer. Innbruck, Wagner.

Krajina, V.J. 1959. Bioclimatic Zones in British Columbia. UBC Botanical Series #1, Vancouver, B.C.

Planck, Max. 1901. On the law of distributon of energy in the normal spectrum. Annalen der Physik Vol. 4, p. 553 et seq.

Odum, H.T. 1971. Environment, Power, and Society. Wiley-Interscience, New York.

Orlóci, L. 1965. The Coastal Western Hemlock Zone on the south-western British Columbia Mainland. Vegetation-environmental patterns and ecosystem classification. In: V.J. Krajina (ed.), Ecology of Western North America. Vol. 1, pp. 18-34.

Orlóci, L. 1971. An information theory model for pattern analysis. Journal of Ecology 59:343-349.

Orlóci, L. 1991. On character-based community analysis: choice, arrangement, comparison. In: Feoli, E. and L. Orlóci (eds.), Computer Assisted Vegetation Analysis, pp. 81-93. Kluwer Academic Publishers, London.

Orlóci, L. 2006. Diversity partitions in 3-way sorting: functions, Venn diagram mappings, typical additive series, and examples. Community Ecology 7:253-259.

Orlóci, L. 2009. Multi-scale trajectory analysis: powerful conceptual tool for understanding ecological change. Frontiers of Biology in China 4: 158-179

Orlóci, L. 2011. Statistical Ecology: a reasoned approach. SSCADA Publishing. Internet Edition: https://www.createspace.com/3476529

Orlóci, L. 2012. Self-organisation and Mediated Transience in Plant Communities. SCADA, London, Canada. Enlarged Online Edition: https://createspace.com/3585127.

Orlóci, L. 2013. On the Energy Structure of Natural vegetation. In search for community governance rules. SCADA Publishing, Canada. Enlarged Online Edition: https://createspace.com/4153484

Orlóci, L. 2015. Diversity analysis, holistic energetics, and

statistics. The resonator complex model in community ecology. EG Report:
https://www.researchgate.net/profile/Laszlo_Orloci/contributions

Orlóci, L. and M. Orlóci. 1985. Comparison of communities without the use of species: model and example. Ann. Bot. (Roma) 43:275-285.

Pillar, De Patta V. and L. Orlóci. 1993. Character-based Vegetation Analysis: the Theory and an Application Program. Ecological Computations Series (ECS): Vol. 5. SPB Academic Publishing bv, The Hague, The Netherlands.

Pillar, V. De Patta and F.L.F Quadros. 1997. Grassland-forest boundaries in Southern Brazil. COENOSES 12: 119-126.

Pillar, V. De Patta, S.C. Müller, Z.M. De Souza Castilhos, A.V. Ávilla Jacques (eds). 2009. Campos Sulinos. Conservação e uso sustentável da biodoversidade. Ministério de Meio Ambiente, Departamento de Conservação da Biodiversidade, Brasilia, DF 70068-900.

Pillar, V. De Patta and L.S. Duarte. 2010. A framework for metacommunity analysis of phylogenetic structure. Ecology Letters 13: 587–596.

Pillar, V. De Patta., C.C. Blanco, S.C. Müller, E.R. Sosinski, F. Joner and L.D.S. Duarte. 2013. Functional redundancy and stability in plant communities. Journal of Vegetation Science 24: 963-974.

Podani, J. 2003. The Evolution and Systematics of Terrestrial Plants. In Magyar. Elte Ötvös Kiadó, Budapest.

Podani, J. 2010. Taxonomy in Evolutionary Perspective. An essay on the relationships between taxonomy and evolutionary theory. Synbiologia Hungarica 6:1-42.

Podani, J. 2015. A Növények evoluciója és osztályozása.(Evolution and classification of plants.) ELTE Eotvös Kiadó, Budapest.

Rényi, A. 1961. On measures of entropy and information. In: PJ. Neyman (ed.), Proceedings of the 4th Berkeley Symposium on Mathematical Statistics and Probability, pp. 547-561. University of California Press, Berkeley.

Revell, L.J., Harmon, L.J. and D.C. Collar. 2008. Phylogenetic Signal, Evolutionary Process, and Rate. Oxford Journals, Life Sciences, Systematic Biology Volume57, Issue 4, pp. 591-601.

Stachowicz, J.J. 2001. Mutualism, facilitation, and the structure of ecological communities. BioScience 51: 235-246.

Stebbins, G. L. 1950. Variation and Evolution in Plants. Columbia University Press, New York.

Sukopp, H. 1987. On the history of plant geography and plant ecology in Berlin. Englera 7: 85-103.

Tilman, D. 2004. Niche tradeoffs, neutrality, and community structure: A stochastic theory of resource competition, invasion, and community assembly. PNAS July 27, 2004 vol. 101, no. 30 10854-10861.

Wildi, O. and M Schütz. 2000. Reconstruction of a 405 yr. recovery process from pasture to forest Community Ecology 1: 25-32.

Wilson, J.B. 2009. Assembly rules in plant communities, pp.130-164. In: E. Weher and P. Keddy (eds.), Ecological Assembly Rules, Cambridge Books Online.

Wilson, J.B., Ulman, I. and P. Bannister. 1996. Do species assemblages recur? Journal of Ecology 84: 471-474.

# *Index*

# Supplementary references

THE VEGETATION PROCESS: A holistic study of long-term community energetics in East Beringia

Authored by Dr Laszlo Orlóci

6" x 9" (15.24 x 22.86 cm)

Black & White on White paper

216 pages

ISBN-13: 978-1499142068 (CreateSpace-Assigned)

ISBN-10: 1499142064

BISAC: Science / Life Sciences / Ecology

ORDER FROM CREATESPACE E-STORE:

https://www.createspace.com/4760258

Process, as the Book uses this term, implies simultaneous execution of two fundamental functions in continuity. One creates complexity, the other reduces it. Ecologists refer to these as community assembly and disassembly. The process requires energy input which determines the momentary potential energy state of the community. This is measurably true in terms of Max Planck's energy-based entropy. We find potential energy increasing when new species (taxa, community elements) are added to or others proliferate in the community, and decreasing when species drop out or their performance declines.

# Quantum analysis of primary succession: The energy structure of a vegetation chronosere in Hawaii Volcanoes National Park

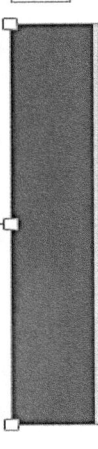

authored by Laszlo Orlóci FRSC

List Price: **$30.00**

**6" x 9"** (15.24 x 22.86 cm)

Black & White on White paper

54 pages

ISBN-13: 978-1492788997 (CreateSpace-Assigned)

ISBN-10: 1492788996

BISAC: Science / Life Sciences / Ecology

The book revisits the classical idea that the potential energy structure of primary succession is a seamless fusion of foot-prints specific to basic processes which operate on all scales – phylogeny, environmental mediation, and chance. The idea is tested in quantum analysis of a vegetation chronosere in Hawai'i Volcanoes National Park. How is the test constructed? What are the outcomes? What do the results tell about primary succession not already known from other sources? Stated in the briefest of terms the test re-quires temporal species performance data...

ORDER FROM CREATESPACE ESTORE:
https://www.createspace.com/4452597

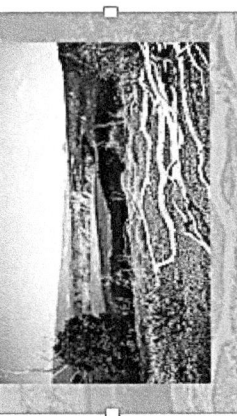

## Quantum ecology: Energy structure and its analysis

Authored by László Orlóci FRSC

List Price: **$30.00**

**6" x 9"** (15.24 x 22.86 cm)
Black & White on White paper
72 pages

ISBN-13: 978-1492183297
ISBN-10: 1492183296
BISAC: Science / Life Sciences / Ecology

Ecology joins forces with quantum theory on the pages of "Quantum Ecology" to create a holistic approach in energy studies.

The infusion of quantum theoretical principles allows the study focus of ecological energetics to shift from the conventional calorific (trophic) flow in ecosystems to the potential energy structure of the vegetation. The books contents cover the theory and techniques in a unique account centred on the energy equation. The equation's component terms define energy footprints specific to ecology's basic processes, such as historic phylogeny, current environmental mediation of transience, and chance. What gives practical value to quantum analysis is its ability to be parameterised by the usual type of survey or experimental data.

The book is offered for classroom use in advanced courses and technical support in research projects.

ORDER FROM CREATESPACE ESTORE:
https://www.createspace.com/4406077

LÁSZLÓ ORLÓCI FRSC

Quantum ecology

Energy structure and its analysis

## Statistical ecology

Like 0

### The quantitative exploration of Nature to reveal the unexpected
**Authored by Laszlo Orlóci Ph.D.**

The book's topics traverse many problem areas in univariate and multivariate data analysis, focussed on current ecological practice. The manner of presentation emphasizes reasoned methodological choices and encourages innovations consistent with the objectives, but mindful of the need to see clearly the regularity conditions which set limits for valid application of statistics in Ecology. The main text is accompanied by external appendices including a technical manual, 47 specialized application programs, and many data files taken from the exercises in the main text. For information please contact: lorloci@uwo.ca

**About the author:**
Orlóci is an INTECOL Distinguished Statistical Ecologist. He is external (academician) Member of the Hungarian Academy of Sciences, and regular (academician) Fellow of the Academy of Sciences of the Royal Society of Canada. He published over 100 papers in scientific journals, numerous monographs and books. His current essays on trajectory analysis, the rules of process governance, and the phylogenetic signal in vegetation transitions have considerable significance for evolutionary ecology and global change science. His present work on energy structures in metacommunities is seminal, pointing to a new direction.

**List Price: $49.90**

Add to Cart

| | |
|---|---|
| **Publication Date:** | Aug 10 2010 |
| **ISBN/EAN13:** | 1453760520 / 9781453760529 |
| **Page Count:** | 372 |
| **Binding Type:** | US Trade Paper |
| **Trim Size:** | 6" x 9" |
| **Language:** | English |
| **Color:** | Black and White |
| **Related Categories:** | Science / Life Sciences / Ecology |

# Statistical multiscaling in dynamic ecology

f Like 0

## Probing the long-term vegetation process for patterns of parameter oscillation

**Authored by László Orlóci Ph.D.**

The Book's conceptualisation of multiscaling theory presents the Next Step in the study of the long-term vegetation process. The context is statistical and the process generating events have proxy in the compositional transitions of the palynological spectra. Familiarity with multiscaling is not a pre-requisite. The reader shall learn from the examples how multiscaling techniques helped to identify the self-similar (fractal) nature of the process, isolate low and high instability phases, locate hotspots of compositional transitions, and link these to delayed climatic effects. He or she shall also learn how to gauge process homeomorphy among sites, isolate the random and directed effects found braided into the process, and do much more within a broad yet formal probabilistic framework. The Book's contents are taken in part from a graduate course offered in the Ecology program at UFRGS in Porto Alegre, Brazil. The examples use palynological spectra from sites on the Hungarian Great Plain and in the adjacent Carpathian Mountains. Application programs are available from the author.

List Price: $30.00

**Add to Cart**

**Publication Date:**   Mar 15 2012
**ISBN/EAN13:**   1475071388 / 9781475071382
**Page Count:**   96
**Binding Type:**   US Trade Paper
**Trim Size:**   6" x 9"
**Language:**   English
**Color:**   Black and White
**Related Categories:** Science / Life Sciences / Ecology

## Self-organization and mediated transience in plant communities

Like 0

## What are the rules?

**Authored by Dr. László Orlóci FRSC**

A novel, signal theoretical solution is sketched out for the ecological problem of how to identify and quantitatively express the assembly rules of plant communities. A case study for testing the solution leads to the astonishing conclusion that the phylogenetic signal outperforms the current environmental signal in intensity close to 7 to 1. This indicates high stability and low inclination to environment mediated transience in the community.

List Price: $25.00

Add to Cart

**Publication Date:** Nov 11 2011
**ISBN/EAN13:** 1461028221 / 9781461028222
**Page Count:** 70
**Binding Type:** US Trade Paper
**Trim Size:** 6" x 9"
**Language:** English
**Color:** Black and White
**Related Categories:** Science / Life Sciences / Ecology

**About the author:**
László Orlóci was born in Hungary in 1932. He holds degrees in forest engineering (DFE Sopron), forestry science and biology (BSF, MSc, PhD University of British Columbia), and DSc h.c. in biology (University of Trieste). Orlóci held appointments as NATO Science Fellow (University College of North Wales), professor (University of Western Ontario), and visiting professor at universities in the Americas, the Pacific, Asia, and Europe. He is an INTECOL Distinguished Statistical Ecologists, external (academician) member of the Hungarian Academy of Sciences, and regular Fellow of the Academy of Sciences of the Royal Society of Canada.

# On the energy structure of natural vegetation

## In search for community governance rules

**Authored by Laszlo Orlóci FRSC**

Briefly about the book ...

Vegetation Science meets quantum theory in the energy-based entropy model of this book. The model is based on Max Planck's postulate that potential energy and entropy are interchangeable parameters in resonator complexes. What is a typical outcome of the model in vegetation studies? The model hands users a set of entropy estimates and probabilities based on which the strength and uniqueness of a metacommunity's energy structure can be characterised in comparative terms.

**About the author:**

Orlóci is an INTECOL Distinguished Statistical Ecologist. He is external (academician) Member of the Hungarian Academy of Sciences, and regular (academician) Fellow of the Academy of Sciences of the Royal Society of Canada. Orlóci published over 100 papers in scientific journals, numerous monographs, books and book chapters. His current essays on trajectory analysis, the rules of process governance, and the phylogenetic signal in vegetation transitions have considerable significance for evolutionary ecology and global change science. His present work on energy structures in metacommunities is seminal, pointing to a new direction.

List Price: $30.00

**Add to Cart**

| | |
|---|---|
| **Publication Date:** | Jan 30 2013 |
| **ISBN/EAN13:** | 1482319373 / 9781482319378 |
| **Page Count:** | 46 |
| **Binding Type:** | US Trade Paper |
| **Trim Size:** | 6" x 9" |
| **Language:** | English |
| **Color:** | Black and White |
| **Related Categories:** | Science / Life Sciences / Ecology |

# Flexible computing in statistical ecology  f Like  0

## External appendix to accompany L. Orlóci's Statistical Ecology
### Authored by Dr. László Orlóci

Problem flexible computing in statistical ecology

The Book describes more than 40 executable (.exe) computer programs and presents examples of application which correspond to the examples included in Statistical Ecology*. The programs are flexibly problem specific and conversational. They allow option-driven selective access to specific statistical tasks. Linkages are possible through standard output and input. The description includes in each case a brief introduction, a record of the start up dialogue, and detailed record input and output sets. The source code is in True Basic. The programs are compiled and linked on a 32 bit Windows XP system and tested up to Windows 7.
The executable program library, data files and a note to users are distributed free of charge to purchasers of the Technical Manual. Requests for download information should be directed to the URL address lorloci@uwo.ca. Prior to running the application programs, installation of a recent version of True Basic (see Internet for sources) on the user's system is strongly advised.
* Orlóci, L. 2010. Statistical Ecology. The quantitative exploration of nature to reveal the unexpected. Scada Publishing, Online Edition. Copies are available from the distributor
https://www.createspace.com/3476529

List Price: $30.00

Add to Cart

Statistical Ecology. A reasoned approach.

| | |
|---|---|
| **Publication Date:** | Apr 05 2011 |
| **ISBN/EAN13:** | 1460972953 / 9781460972953 |
| **Page Count:** | 142 |
| **Binding Type:** | US Trade Paper |
| **Trim Size:** | 6" x 9" |
| **Language:** | English |
| **Color:** | Black and White |
| **Related Categories:** | Science / Life Sciences / Ecology |

# *Reader's notes*

# *Biographic notes*

 László Orlóci was born into a military family in Hungary in 1932. He graduated from high school summa cum laude in Forestry Science; holds university degrees in forest engineering (DFE Sopron), forestry science and biology (BSF, MSc, PhD University of British Columbia) and DSc *h.c.* in biology (University of Trieste). He held appointments as NATO Science Fellow (University College of North Wales), professor (University of Western Ontario), emeritus professor (Western University) and visiting professor at universities in the Americas, the Pacific, Asia, and Europe.

Orlóci is INTECOL's Distinguished Statistical Ecologist; external (academician) Member of the Hungarian Academy of Sciences; regular (academician) Fellow of the Canadian Academy of Sciences (of the Royal Society of Canada).

Orlóci published over 100 papers in scientific journals, numerous monographs, books and book chapters. His current essays treating trajectory analysis, the rules of process governance, the phylogenetic signal in vegetation transitions, and quantum analysis have considerable significance for evolutionary ecology and global change science. His present work on the energy structure of metacommunities is pointing to a new direction in ecological energetics.

Orlóci is married to author, forest engineer Márta Mihály, Sopron forest engineering alumna. Their daughter Martha Barbara is a Geography graduate of Western University. They have two granddaughters, Kathryn and Ruth Orlóci-Goodison. Kathryn is enrolled in Natural Resources Management at Lake Head University. Ruth is 1st year Presidential Scholar in Biology at the same university.